Modern Physics for Beginners

Duality, Atoms, Nuclei, Relativity and Universe

CONTENTS

Preface	3
Basics	6
Dual nature of Light	8
Dual Nature of Matter	20
Electromagnetism	26
Atomic Structure	32
Nucleus	45
Relativity	54
Our Expanding Universe	66
Applications	72
Questions	77
Glossary	86
About the Author	89

Preface

Physics of the 20th century is often called modern physics. Before that, we were just trying to understand the basics. In the 20th century several important discoveries and theories were put forward – discovery of electron and internal structure of atom, Einstein's relativity and quantum theory to name a few. Most ideas were beyond human comprehension, some of the theories were bizarre!
But today we use advanced technologies like GPS, Microcomputers, Plasma TVs, and Mobile Phones – all of these because of the physics of the 20th century. This book is a glimpse of modern physics. Anybody who is a passionate physics enthusiast can read this book, call it the layman's guide to modern physics – short and sweet.

In this book, we will look at:

1. Wave-particle duality
2. Photoelectric Effect
3. Discovery of electron
4. Atomic models
5. Nuclear forces
6. Energy-mass equivalence
7. Special relativity
8. General relativity and
9. Modern applications of modern physics

Applications of modern physics that have been included in this book are TV, Fire alarm, electron microscope, nuclear reactors, atomic/nuclear weapons, solar panels, GPS and even time machine (theoretical) hoping that this will further help the reader in their understanding of the concepts.

One more important thing: This book is a simple beginner's guide to modern physics with simplified explanations, appropriate diagrams, handy illustrations and related stories so that you do not get bored as you discover the complex theories of physics. Happy reading!

Basics

Before we study modern physics we will quickly recap the basics:

1. Newton's three laws of motion describe how objects respond when forces are applied on them – First law of motion defines force, second law of motion tells that objects accelerate when acted upon by unbalanced forces, third law of motion is about action and reaction

2. In circular motion, the direction of the revolving object keeps changing as it moves along the circle. This means that it is under acceleration since change in direction corresponds to change in velocity

3. Energy is the ability to do work. Work, in the language of physics, is the displacement caused by a force. An object with more energy can do more work, i.e. cause more displacement

4. Energy is conserved. But one form of energy can be converted to another. For example, sunlight can be converted into food!

5. Matter is made up of atoms. There are different types of atoms in the universe

6. There is no such thing as absolute motion. An object may appear to be at rest with respect to your perspective and in motion with respect to someone else's

Dual nature of Light

This story is very old. We always see colorful things around us but we hardly question what they really are. Can we imagine life in black and white? And what exactly is light!? The first person who tried to understand the behavior of light through experiments was an English scientist – Isaac Newton was his name! With his remarkable genius, Newton understood that white light of the sun was actually made up of seven different colors!

To demonstrate his idea, Newton painted a circular disc with seven different colors. Newton's disc is a simple activity that anybody can do at home. When you rotate this disc about an axis (say a pencil) you will notice that the seven colors blend almost perfectly and the disc will appear dull white in color!

Newton supported another interesting idea that light was made up of tiny particles. In fact, all those colors were nothing but different arrangements of vivid particles. Newton strongly promoted this particle theory of light. He could therefore explain the reflection of light in terms of particles of light bouncing off the mirror. In this way, Newton strengthened the particle theory of light...

But another great scientist, Christian Huygens believed in the wave theory of light. Wave theory of light was a different clan which supposed that light was nothing but a wave. Just like ripples of water can be considered as water waves, light could also be thought of as consecutive waves traveling in some medium.

A source of light emits a wave in all directions. Each point on this 'primary' wave serves as a new source for a 'secondary' wave. This new secondary wave is fainter than the older one. In this way, waves after waves are generated in the medium:

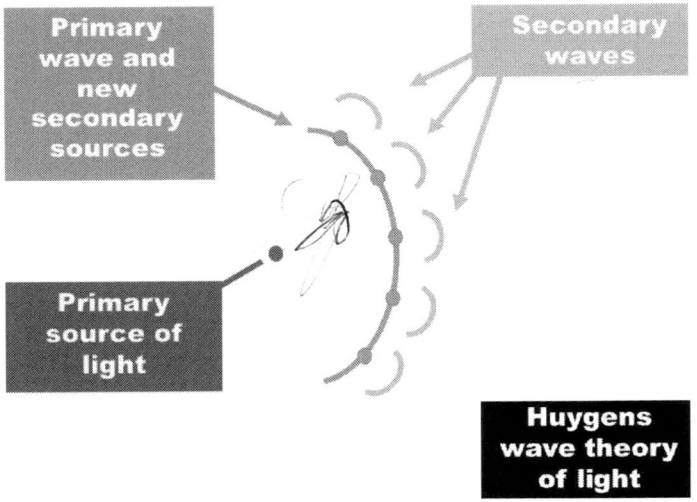

This allowed Huygens to explain not only reflection but also the phenomenon of refraction. Unfortunately, Huygens died early and his ideas were overshadowed by Newton's domination in the world of physics. The wave theory of light was forgotten until...

Until someone actually performed experiments to prove the validity of wave nature of light. To visualize a light wave, we use the model of crest (mountain tops) and troughs (valleys) which travel one after another and carry energy. The reason why we use this repeated pattern as shown in the image below is because these waves are generated one after another. The waves often represent sinusoid (or trigonometric sine).

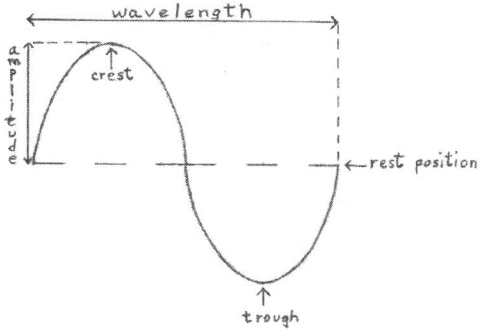

Thomas Young, an English scientist, knew that sound was also a wave. One important thing that he recognized was the fact that mixing of different sounds could sometimes result in complete silence! He imagined the two sound waves as sinusoids interacting with each other...

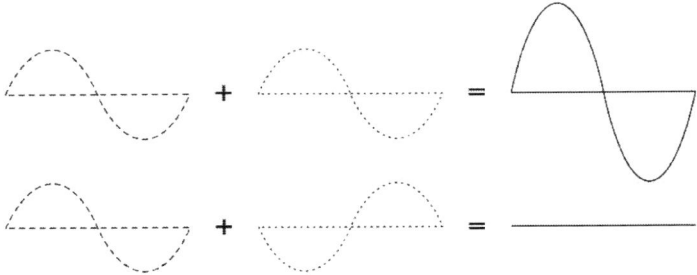

The crest of one wave when mixed with the trough of another, produced nothing! He wanted to repeat the same experience with light. His success depended on whether or not light was a wave. Young's experiment required special apparatus and a very careful design. The results were amazing! At specific conditions, he was able to 'see' the darkness when light mixed with light! This was perhaps the most important experiment ever performed!

Interference Pattern

This is Young's interference pattern which shows bright and dark areas. These almost rectangular patches repeat one after another. In the dark region, there is destructive mixing – troughs cancelling crests. On the other hand, when troughs coincide with troughs and crests coincide with crests, there is constructive mixing – the bright region. The energy emitted out of the two light sources is redistributed to create this pattern. The wave theory has finally won!

Wait! The story is still incomplete. Although the particle theory of light could not explain this interference thing, but there were still other phenomena that wave theory could not explain too! Photoelectric effect is what we are talking about...

Heinrich Hertz, a German experimental physicist once discovered that whenever light rays of suitable energy were incident on a metallic sheet, electrons were emitted out of that metal. What was happening? How could an incident ray of light trigger an electron out of the metal? Photoelectric effect became a riddle! And since wave theory supporters could not come up with an explanation to what was happening at the atomic scale, particle theory was the only hope.

Max Planck, another German physicist published an important idea that radiation (light) was made up of discrete particles. The smallest unit of energy was called quantum. He said that energy of any quantum was directly related to the frequency of the light source. For example blue light has higher frequency than red light, thus he said that blue light particle will have higher energy than the red light particle. These particles were later named **photons.**

Albert Einstein, yet another German physicist, utilized Planck's theory to explain the photoelectric effect. He also supported the idea that light was made of particles (photons) which somehow caused electrons to get out of the metal.

Since each photon had energy, it must have collided with an electron and transferred some of its energy to the electron. This idea can be understood with the help of billiards.

When we hit a ball with a stick, we give it the energy to move, in other words, we give it the necessary energy to strike and disturb all the other balls. Similarly, photons have the ability to disturb the electrons of an atom – Einstein's idea is simple to understand...Photoelectric effect is a type of particle-particle collision.

Wave theory could not do it, and therefore Einstein's explanation was accepted – he even received a Nobel Prize for the same theory!

But there was a problem – some of the phenomena could be explained by wave theory while others were only understandable with the particle theory of light. Which theory was right? Both seemed reasonable...

It was settled that light was of dual character. With our limited minds we will never be able to understand the true character of light. Sometimes it will behave like a particle and sometimes it will behave as if it were a wave. At any time, it will show both the characters equally. This is called the **wave-particle duality**.

Now let us talk about an important application of photoelectric effect – smoke detector alarms! An invisible electromagnetic radiation (beam of photons) is continuously falling on a detector (metallic plate) causing the electrons to come out and form a conducting stream (electric current).

Sudden smoke disrupts the path of the beam of photons and the detector is so accurate that even minimal disturbance is enough to trigger an alarm. Another application of photoelectric effect is the use of metallic sheets in solar panels. Sunlight is used to 'trigger' the electrons to maintain a stream of current. This is used to manufacture electricity at home!

Summary

Character of light was always a topic of discussion. Sometimes the wave character dominated and sometimes the particle character tasted victory. It was finally settled that light was a **"wavicle"**

Dual Nature of Matter

Einstein published another important paper in the early 20th century which proposed that matter and light (energy) were equivalent. Mass was just a compact form of energy, he said. Energy-mass equivalence formula was born!

If matter (mass) and energy (light) are similar, does it mean matter should behave the same as light? Should matter particles (like electron) also show wave-particle duality?

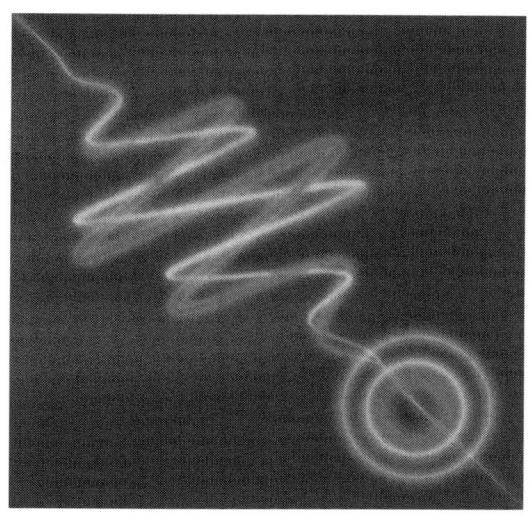

Electrons, protons and neutrons are particles that we know about very clearly. These are sub-atomic particles. Are they in fact energy carrying waves? But if they are waves then why aren't all the things wave-like? All these things around us are made up of sub-atomic particles! Can we imagine these particles like waves? And what is this wave we are talking about? Can we think of the physical appearance of a wave? How does a wave look like?

Let us first know what a wave is. This time we will look deeper. In the previous chapter, I just told you the way we represent waves on paper and in equations (sinusoids). What a waste! We should be able to visualize waves in our heads! We should know what they really are...

Both, waves and particles, can carry energy. But they differ in the way they do it. A particle is confined like a dot. On the other hand, a wave is spread out like a sheet of paper, its energy is not confined in a single location. This is the major difference between the two.

Louis Broglie, a French physicist claimed that since light and matter were symmetrical in nature, particles should also behave like waves. What does that even mean? He meant that particles should not be confined in one specific space but they must behave like waves, spread out and distributed...

That remained only an idea until Davison and Germer – American experimental physicists performed experiments that confirmed Broglie's idea. They studied electrons and observed that electrons behaved like waves and produced interference pattern just like usual light waves! Following is the image of electron interference pattern *(d and e)*:

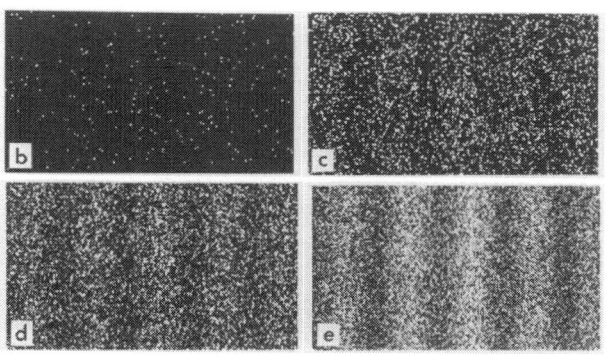

After this experiment, Louis Broglie was awarded the Nobel Prize in physics. This was wild! Particles were also like waves! But why don't we observe the things around us to behave wave-like? Why are they not spread out? That is because, wave nature of matter is only measurable for small particles such as electrons.

The wave nature of any object decreases with increase in its momentum. Momentum is a product of object's mass and velocity. So it turns out that only the moving particles have wave-like characteristics. Objects around us are so big and massive that their wave nature is negligible (nothing).

On the other hand, wave nature of electrons inside an atom is very significant. In fact it is so much noteworthy that one cannot even find an electron inside an atom! This we will discuss in the chapter of Atoms!

Summary

Matter is also wave-like! But wave nature of matter is only significant for small particles and not for the objects we see with naked eyes. Also, wave-particle

*duality is the harsh
reality of nature...*

Electromagnetism

Before we study atomic structure, we need to study about electric charges and fields. This is because inside atoms are tiny charged particles like electrons and protons. What is charge? How do we know whether a particle has charge or not? Do all particles have charge?

The idea of charge is very old. Benjamin Franklin first performed experiments with electric charges, although he did not think of them as charged particles, he thought of electricity as a result of moving electrical fluid. He discovered that the electrical fluid could move from one body to another upon rubbing...

Just like mass is a property of a particle, electric charge is also a property. But not all particles have electric charges. For example neutron is a neutral particle. To know whether a particle has electric charge or not, we perform experiments.

A battery has two ends – a positive end and a negative end. Negative charged particles like to go towards the positive end of the battery. On the other hand positive charged particles like to go to the negative end. By looking at where the particle goes we can deduce what type of charge it has. A neutral particle won't go anywhere!

Electric charge also gives a particle the ability to have its own electric field. What is a field? A field is an area within which the presence of the charge can be 'felt'. All charged particles have fields. If the magnitude of charge is more, the field is large and strong. There are two types of charges – positive and negative.

Negative and negative/positive and positive repel each other. On the other hand, positive and negative charges attract each other. Therefore whenever one charge is in the vicinity of the other, it feels the presence of the second charge because of its field.

All charges at rest possess an electric field. This is called static electricity. However, when electric charges start moving with constant velocity they make a stream known as 'electric current' which carries electrical energy. This flow of charges also creates a magnetic field in the vicinity. This fact was first discovered by Hans Oersted.

One day, Hans Oersted was performing experiments with electric wires. He accidentally discovered that whenever he switched on the current the nearby magnetic compass started acting weirdly. A magnetic compass has a magnetic needle which moves only when there is a magnet nearby.

There was no magnet! There were only wires carrying electric current! Oersted concluded that the moving charges or electric current was creating a *magnetic field* around the wire with which the magnetic needle of the compass was interacting...

Soon scientists realized that electricity and magnetism were co-related. One could produce another! This is called electromagnetism. In the first chapter we studied about light – light is actually an electromagnetic wave. This means that what we see as visible light (color) is actually a combination of electric and magnetic fields.

Colors of a rainbow are also electromagnetic! These are different electromagnetic waves that differ in frequencies and energies...

Charges at rest possess static electric field. Charges in uniform motion (constant velocity) create current which in turn produces a magnetic field. What do accelerating charges create? In the late 19th century, Heinrich Rudolph Hertz, a German experimental physicist was working with accelerating charges. He finally produced radio waves with those accelerating charges! Radio waves are long wavelength waves. He realized that with varying the acceleration of the charges, he could create various electromagnetic radiations!

Numerous new electromagnetic radiations were discovered and produced with the help of accelerating electric charges. X-rays, Microwaves and infrared waves were a few of them – they find use in hospitals, kitchens and our remote controls!

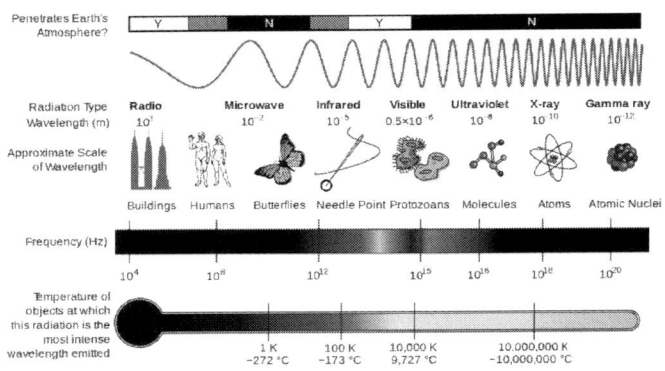

Atomic Structure

Scientists have always tried to understand the shape and structure of an atom. Long time ago, people considered that all objects were made of indivisible things called *atoms.* But in the early 19th century sub-atomic particles – particles smaller than the known atoms – were discovered! Thus everyone wanted to develop models to explain atomic structure...

Joseph Thomson discovered the electron. An electron is a charged particle. Thomson was working with cathode ray tubes, the ones that are now used in CRT TVs. A CRT TV uses cathode ray tubes with a fluorescent screen at one of the ends. When electron beam hits the screen, flashes of light are produced which we see as picture.

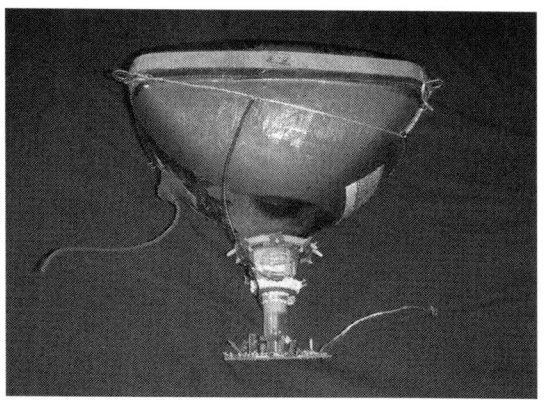

These glass tubes have electrodes (metallic plates) fixed on the ends. Appropriate conditions are maintained and high voltage is applied across the electrodes (just like you connect the two ends of a battery)

When Thomson used a cathode ray tube, he observed a sharp beam between the two electrodes. Thomson applied an *external* electric field and observed that the beam deflected towards the positive end of the external electric field as shown below:

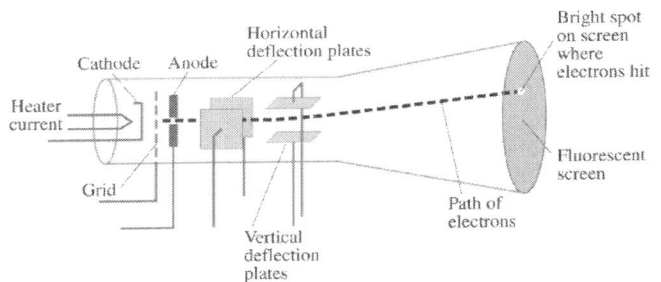

Thomson deduced that the beam must be made of negatively charged particles. Electron was finally discovered. But wait! There is more...

Since things around us were neutral in general, Thomson realized that the atom must also contain a positively charged particle to balance the negativity of the electron. This particle was discovered later and was named proton.

Proton was found to be much heavier than the electron because, the beam made up of protons was very hard to deflect (very strong electric fields were used!)

Finally Thomson wanted to describe what an atom looked like on the inside. For that he had to use his imagination. Since sphere was the most **stable** object (because all water droplets were spherical, planets and stars were spherical) he deduced that atom was also spherical in shape...

His model is often called the **plum-pudding model.** In this model, electrons are embedded inside a positive sphere. The mass of the atom is uniformly distributed throughout the sphere. This model is shown in the image below:

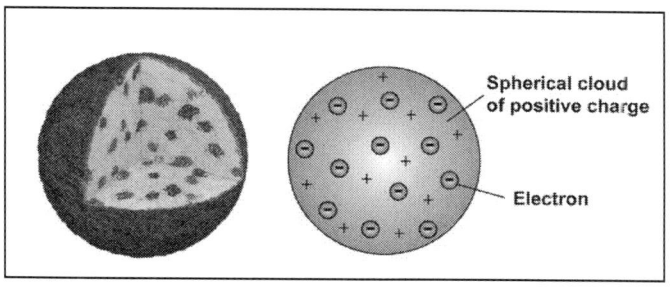

Nothing is perfect. In physics, theories are constantly being challenged and updated. Thomson's model of atom was also tested with experiments. Rutherford, a New-Zealander, was performing advanced experiments with positively charged alpha particles and gold atoms.

With the help of his students, Rutherford designed an alpha particle experiment in which a beam of high energy alpha particles would collide with gold atoms kept at a distance. The result of the collision would be observed on a screen.

If Thomson's theory was correct, any colliding alpha particle would bounce back after the collision. It is obvious that when you throw a ball towards a solid wall, it bounces back at you! According to Thomson's model, atom is a dense sphere but Rutherford's experiment produced one shocking result: **Most alpha particles passed straight through the gold atoms – as if there was nothing blocking their way!**

Rutherford concluded that Thomson's model was not the true picture of an atom. His experiment suggested that most of the atom was empty space! However, he also found that a few alpha particles were deflected at small angles (just like a curving ball)

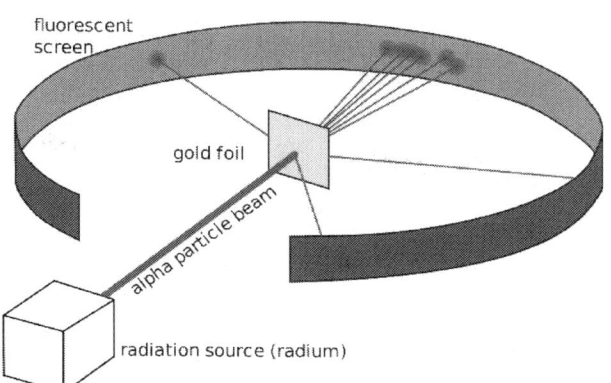

Another important observation was that, 1 in 20,000 alpha particles (a very small number), rebounded, which suggested that the alpha particle hit something really hard (dense) and bounced back! Rutherford collected all the results and created his own picture of atom. His model is often called the **nuclear model** of atom. In this model there is a very small space at the center of an atom within which all of the mass of the atom is confined. This is called the dense nucleus. Around the nucleus there are small and light-weight electrons that move in their fixed orbits – just like planets revolving around the sun!

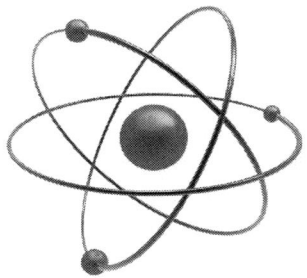

A small ball of protons around which tiny electrons circle with love – interesting! The reason for this strange love is their electric field interaction. Proton and electron attract each other as they are oppositely charged particles.

To calculate mass of an atom on paper, Rutherford added individual masses of protons inside the nucleus. He did not take electrons into account because they were extremely light weight and thus, negligible. However, when other scientists performed experiments to determine atomic masses more accurately, they found different results – their measurement was more than what was calculated by Rutherford!

Therefore, Rutherford predicted the existence of a new sub-atomic particle inside the nucleus which he called the neutron. And it had to be neutral – otherwise it would disturb the neutrality of an atom! With this prediction, Rutherford received the Nobel Prize in Physics. One of his students James Chadwick, an English physicist, discovered the neutron for real! James observed neutral radiation coming out of several atoms when subjected to appropriate conditions. The neutral radiation that he observed must be composed of neutral particles – exactly what Rutherford had predicted! Chadwick also found that the mass of the neutron was almost the same as that of the proton.

Now atom has a beautiful family:
1. Electrons
2. Protons
3. Neutrons

However physics is always about constant development. Even Rutherford's model was not perfect. Changes were necessary. What was the problem with his model?

Whenever a particle moves in a circular orbit, it is continuously being accelerated. This is because at each and every point of the circular motion, direction of the moving particle changes. This change in direction corresponds to the changing velocity, which in turn, corresponds to acceleration.

So inside these atoms, electrons are under constant acceleration. But according to Maxwell's electromagnetic theory, accelerating charges emit radiation and lose energy, because radiation, is a form of energy and energy, needs to get out of somewhere! By the law of conservation of energy, electron's kinetic energy (energy by virtue of its motion) should get converted into radiant energy. However this is not observed in real life. Your body is not emitting any kind of light radiation right now! Your body is made up of atoms, remember!?

If however, the kinetic energy is getting converted into radiation, the electron must slow down eventually. And the positively charged nucleus will pull the light-weight electron towards itself – and the electron will eventually collide with the nucleus! This Rutherford picture of atom is horribly wrong! Atom is not unstable like this!

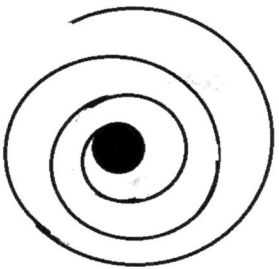

The orbiting electron will emit energy and thus, lose its kinetic energy. It will slow down and collide with the nucleus.

How do we correct this problem? Niels Bohr, Danish physicist, came to the rescue. He used Max Planck's quantum theory and stated that electrons revolved within **quantized** orbits. Orbits are not randomly placed around the nucleus as Rutherford may have assumed. "They are there for a reason", Bohr said. These orbits can be thought of as 'fixed' energy states. Electrons while revolving in these fixed states cannot emit any kind of radiation. These fixed states can be calculated with mathematics. But we will not go into that right now...

"If however, an electron absorbs energy from any outside source", Bohr said, "It will climb up the energy states". This means that all around the nucleus there are 'fixed and unique' energy states and the further away you are from the nucleus, the more energy you possess. But since electrons are happy only in their respective states – blame the natural laws – they want to jump back to their 'ground state'. This is all set up by nature – we develop theories thinking that nature works the same way. If the theory does not fit previous observations we make another theory and so on...

So Bohr's theory said that orbits around the atom were not arbitrary but stationary energy states. This model was accepted over Rutherford's model of atom because it provided a better explanation! Rutherford's model of atom was simple but it was unstable. On the other hand Bohr made only one change – that of fixed energy states. Everything else remains – atom is mostly empty, nucleus is small and dense, etc.

But science is all about development...

Once again atomic structure needed change. Even Bohr's model of atom was not satisfactory! The new model of atom incorporates dual nature of matter and uncertainty principle. However we won't look any further. In the book 'autobiography of an atom' we will look at the funny side of an atom telling us stories about his life and his family. I am working on that book right now!

Summary

Atom is a tiny particle. But even atom has an internal structure! Within an atom is a universe of empty space with few electrons revolving around a dense nucleus. However in physics, we keep on developing our theories when we learn of something new. This is often called evolution...

Nucleus

In the previous chapter, we looked inside an atom and discovered that almost all of the mass of the atom is confined in an extremely small space called the nucleus. Since the nuclear volume is so small, its density is enormously high!

Rutherford's model of atom suggested that nucleus was composed of neutrons and protons. The mass of the atom is almost equivalent to the sum of masses of all the protons and all the neutrons in the atom. Neutron has no charge and proton is positively charged particle. So overall the nucleus is positively charged 'sphere'.

The first thing when you study about electric forces is the fact that like charges repel. Moreover, the electric force gets stronger when the distance between the two charges gets smaller. Inside the nucleus, protons – positively charged entities – live freely as if they do not feel the electric force of repulsion!

The nucleus is so small and therefore, the electric force would be tremendously high inside the nucleus! But there must be something that is keeping these protons together. That something must hold the protons and neutrons together...

It was concluded that nucleus was home to a 'mysterious' binding force stronger than the electric repulsive force. Only then, the nucleus will remain stable. This force is called the nuclear force. Nuclear force is responsible for sticking the protons together. It is the glue of the nucleus!

Mass of the neutron is almost similar to that of the proton. Neutron is only a "little" more massive. The mass of nucleus is often called the atomic mass (since electrons outside are negligible) and the number of protons in an atom is called atom's atomic number.

Experiments in chemistry in the 20th century suggested that radiations were emitted from within the nucleus. Nobody could explain the source of those energy radiations except Albert Einstein. In the early 20th century Einstein published a ground-breaking scientific paper stating that matter and energy were equivalent. This is often called energy-mass equivalence.

According to this rule, mass and energy are equivalent. Einstein said that the nuclear energy was originating from the nuclear mass i.e. whenever a burst of nuclear energy occurred, the nucleus lost some of its original mass.

Energy-mass equivalence also states that 'mass-energy' emitted by the nucleus is equal to the product of "mass lost" and square of the speed of light. This is enormous amount of energy because the speed of light is very high! More experiments were conducted and more radiations were observed.

What makes a certain nucleus emit radiation? Why would a nucleus want to lose its own mass!? There is something we have to study before we get the answer – binding energy!

Who is providing the nuclear force? For example, because of the electric potential energy there is electric force between charges. So what is it exactly that is responsible for the nuclear force? This type of energy is called binding energy because, it binds the nucleus together.

However there are many different kinds of atoms and so, different nuclei must have different binding energies. Binding energy holds the repulsive protons together – it is obvious that a more massive nucleus would require more binding energy to sustain its structure. Therefore nuclei that have low binding energies are easier for us to break apart...

If we hit a very heavy nucleus with a speeding particle (such as neutron), it will further destabilize the nucleus. This means that the binding energy will be unable to keep the nucleus together. As a result the heavy nucleus will be forced to emit a few particles out, so as to become a stable structure again.

For example, Uranium nucleus might emit alpha particle out of it when hit by a speeding particle. This emission of 'special particles' is called nuclear decay.

Let's get back to binding energies. Nuclei with low binding energy would like to get stronger, they would be more than happy to do so! They achieve this either by fusion or fission. We need to look at the curve of binding energy of different nuclei:

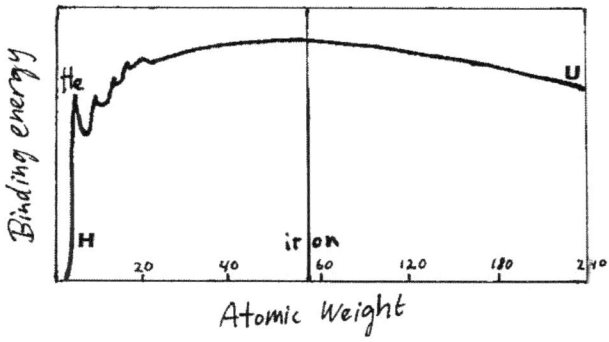

As you can see in the picture above, elements like hydrogen and helium have really low binding energies – they are at the bottom of the curve on the left. As we go further along the curve, the binding energy witnesses a significant rise. After this rise, there is again a fall in binding energy – look at Uranium!

Uranium nucleus has 235 nucleons (no. of neutrons + no. of protons) but its binding energy is lower than lighter nuclei like Iron! This makes it so unstable!

So nuclei with unusually low binding energies want to get more stable. Fusion is a process in which two light nuclei mix or combine together to form a heavier nuclei. In doing so there is a gain in binding energy. For example two hydrogen atoms fuse to form helium – look at the binding energy curve again – helium has much higher binding energy than hydrogen.

On the other hand nucleus of Uranium is so heavy that it can hardly fuse with others! It cannot fuse and form yet another heavy nucleus! So what does it do? Fission is a process in which a heavy nucleus breaks down into lighter nuclei. This will lead to stability and happiness...

These two processes are very popular in science and technology.
1. Fusion is used in stars to produce natural light
2. Fission is used in nuclear power plants to produce electrical energy

Stars are mainly made up of hydrogen. Hydrogen is their fuel. They burn hydrogen into helium and radiant energy, which we see as light. In this way they are converting their mass (hydrogen) into energy (light) – Einstein's energy mass equivalence once again!

The sun burns 600 million tons of hydrogen every second.

On the other hand, nuclear reactors use Uranium as fuel to create electricity from fission energy. Uranium is called fission material. When it disintegrates into smaller nuclei it emits energy as well! However nuclear reactors also come with a threat – nuclear accidents are very dangerous as we all know...

Nucleus is a tremendous source of energy. It is so small but it is capable of destroying planets! Yes, you heard (read) that right! Nuclear weapons use energy-mass equivalence and convert all of their mass into destructive energy. Hiroshima and Nagasaki are the only cities that have faced the wrath of nuclear weapons.

I wish there are no more wars. I wish there are no more nuclear weapons.

Getting political all of a sudden!

Summary

Nucleus is very small. And it is also very dense! Protons inside the nucleus are held together by nuclear force (or binding energy), however some nuclei are unstable and they either fuse

*or break apart to
gain stability.*

Relativity

What is motion?

That book on your table may seem stationary to you but for someone on the moon, it is moving at a tremendous speed, since earth is circling around the sun! Motion is all a matter of perspective...

This means that there is nothing as such as 'absolute' motion or 'absolute' rest. One observer may see an object to be moving and another can see the same object to be at rest! For example, passengers on a bus seem to be stationary with respect to each other. But someone on the outside clearly sees them moving ahead!

This is called relativity. However this notion of relativity gets very ugly when we talk about 'light'. Light is an electromagnetic wave as we already know, and it travels with an amazing speed of 300,000 kilometers every second!

Albert Einstein published an important result in 1905, early 20th century. In fact in 1896, nine years ago, Einstein reported:

> *"If I pursue a beam of light with the velocity of light, I should observe such a beam of light at rest. However, there seems to be no such thing, whether on the basis of experience or according to Maxwell's equations. From the*

very beginning it appeared to me intuitively clear that, judged from the standpoint of such an observer, everything would have to happen according to the same laws as for an observer who, relative to the earth, was at rest. For how, otherwise, should the first observer know, i.e., be able to determine, that he is in a state of fast uniform motion? One sees that in this paradox the germ of the special relativity theory is already contained. Today everyone knows, of

> *course, that all attempts to clarify this paradox satisfactorily were condemned to failure as long as the axiom of the absolute character of time, viz., of a simultaneous, unrecognizably was anchored in the unconscious. Clearly to recognize this axiom and its arbitrary character really implies already the solution to the problem."*

Okay! So what exactly is he trying to say?

Suppose you travel at the speed of light – then according to the so-called "usual" relativity, you must observe another beam of light to be at rest! However this is not the case. Einstein published an important result, "The speed of light is same for any observer"

Suppose there is an electric bulb fixed at position 'X' in a very large room. You start moving from the same place, 'X', at the speed of light in the same room. Just after you start moving, the electric bulb is switched on by someone else. If the usual relativity applied, the light from the source will not be able to catch up with you. You will always be ahead! This also means that the large room will forever remain dark 'for you' as you continue moving, even when the bulb is switched on! But if the room is dark how will you be able to keep moving!? This is not acceptable in real life. This is a strange paradox.

Einstein's idea was that the speed of light was measured same by anyone who observed it/ Einstein 'created' a theory in which he could conserve this idea. In physics we publish our ideas which we truly believe in – Einstein believed in his intuitive power. However every theory is subjected to tests and fortunately for Einstein, tests favored his theory!

Postulates of Special Relativity:
1. The laws of motion are same in every 'inertial' frame of reference. An inertial frame is non-accelerating reference frame

2. The speed of light is the same independent of the observer
3. Speed of light is the cosmic speed limit

Physical laws must be the same for anyone, anywhere in the universe. Also the speed of light is the cosmic speed limit. No one can go faster than the speed of light. This rule is as natural as the rule of conservation of energy. What will happen if someone tries to go faster than the speed of light? Given that this is a fundamental law, how will nature prevent its violation?

Einstein spent months thinking about the same – he finally used energy-mass equivalence to derive the answer to this problem. Suppose someone approaches very close to the speed of light. In order to increase its speed further, it will need to accelerate, but nature does not like this!

Nature knows that a more massive body is more difficult to accelerate than a lighter body – Newton's second law of motion! Thus, nature would prevent the violation of cosmic speed limit by making that "someone" fat! As he/she tries to accelerate, more and more of its kinetic energy of motion will turn into fat! He/she will become heavier and heavier – so much that they will not be able to accelerate any further! This is an idea, of course. An idea which in principle should work, since energy-mass equivalence is true and the second postulate is also true.

Special theory of relativity says that quantities like time are a matter of perspective – for example – two twin brothers separate one day, to test this premise. One of them leaves on a long space tour while the other stays back on earth. And when they meet again, the one who went to space will be younger. This is called Twin Paradox!

This difference because the two brothers measured their own personal times! The one who remained on earth spent the usual earth days. On the other hand, the one who went through space at amazing speeds experienced his own time – much slower – and returned younger.

(Anti-aging business)!

Exactly 10 years later in 1915, Einstein published another 'version' of relativity – this time explaining the 'force' of gravity. This theory was called the general relativity.

General theory of relativity is not only a theory about gravity but also a theory about space, time and light! This is illustrated in the figure below:

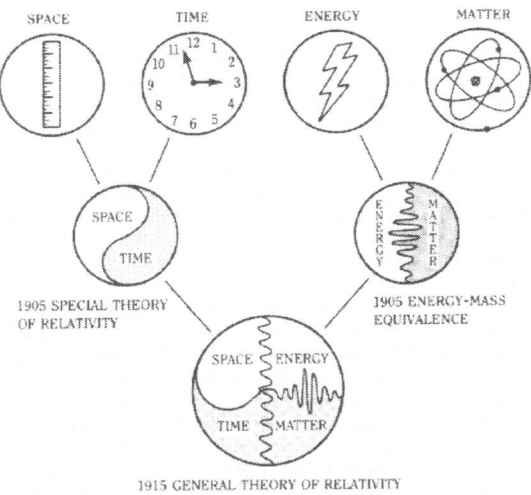

Einstein's special relativity was a combination/unification of space and time. And his energy-mass equivalence unified energy and mass. When these two different unifications are further unified into one big product, we get general relativity!

Let us talk about general theory of gravity by Albert Einstein. Newton visualized gravity as a mere force between two masses separated by a distance. But Albert Einstein fully understood the force of gravity. He pictured gravity as ripples/waves traveling through space and time.

Suppose the sun suddenly disappears so that the planets no longer feel its gravity – according to Newton, all the planets will instantaneously fly away from their orbits. However Einstein said that gravity was like a 'signal' from the sun. If nothing could go faster than the speed of light then how could the planets instantaneously receive the gravity signal? Sunlight takes 8 long minutes to reach us here on Earth! Then how will Earth instantaneously fly out of its orbit?

Newton's theory had a problem. Newton supposed gravity to be a force capable of notching infinite speeds! But that was not the case – gravity was not an instantaneous force. Einstein was right. Only recently in 2016 exactly 100 years since he published general relativity, scientists have finally detected gravitational wave signals from a pair of colliding black holes!
Okay!

But general relativity was also a theory about light and time. According to this aspect of the theory, time and light are affected by the force of gravity. This means that time would slow down near massive objects such as planet earth.

Time is elastic.

This strengthens the concept of a machine which could manipulate time – time machine! Will we be able to create such a machine is a very difficult question to answer. But not only time, gravity also affects light! Light is bent by a gravitational field. This strengthens the concept of a body so massive such that it bends light so much that no one could find it – black hole! Black hole is a dead star whose gravitational field is so effective that not even light can escape from its grasp! Light is the fastest, remember?

Summary

Einstein's theory of relativity can be broken down into two aspects –

special and general relativity. Special relativity is more about position and time. On the other hand general relativity is about gravity.

Our Expanding Universe

An air-filled balloon is static. Static means that it is not changing – expanding or contracting – with time. However if you fill more air or if you take out air, you either expand it or contract it. This is an example of a changing universe. The universe we live in is expanding, but how did we discover this amazing thing!?

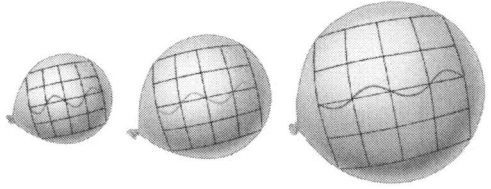

Our understanding of the universe has significantly increased decades after decades, centuries after centuries. Aristotle, a Greek philosopher once claimed that earth was a flat plate and that all the stars revolved around that plate. Today we know that our planet is a tiny, a very tiny spherical pea-sized structure in front of the vastness of the universe!

Today we also know that our earth is revolving around the sun and that sun is also revolving around the Milky Way galaxy. The galaxy is also moving! Much of our understanding of the universe has come from observation of stars.

For example by looking at the brightness of stars we can tell how far they are. We can compare the brightness of the two stars and determine which of them is farther away. We can also look at stars' colors and determine their body temperature! Yellow stars (like our sun) are the coolest. Not only temperature, we can also tell what the star is made up of, just by observing its starlight. In this way we have found that most chemical elements formed inside the cores of early stars...

By observing the starlight, we can also tell whether the star is static or moving! In fact if we observe the entire galaxy this way – we can tell whether it is moving away or coming closer! This is exactly what we did, to determine that the universe was expanding with time – like a balloon!

Edwin Hubble, an American astronomer was studying the light of different galaxies in the universe. Almost all the galaxies that he observed appeared reddish. From this observation, he concluded that all the galaxies were moving away from us. He suggested that the universe was expanding!

This came as a surprise for those who supported the static-universe model. Einstein was one of them! Albert Einstein was attempting to make the complete model of the universe by using general relativity but he could not succeed because he presumed the universe to be static!

How can we tell whether the universe is expanding or contracting from observing the light of distant galaxies? For this, we will need to understand the Doppler Effect. You may have heard of this before! Remember Sheldon Cooper? This is exactly the same principle except that this time we will apply it to light waves.

Doppler Effect!

For sound waves, Doppler Effect is an easy observation. We can tell that a car is getting closer and closer just by hearing its sound. Its engine sound keeps on changing – this we can tell, because of the Doppler Effect. We can also tell if a car is getting away from us just by hearing its sound fade away with time.

What is happening with sound?

In the case of sound waves, as the car comes closer, the frequency of the sound increases and we perceive this as decreasing distance. As the car goes away, the frequency reaching us drops continuously letting us know that the car is gone!

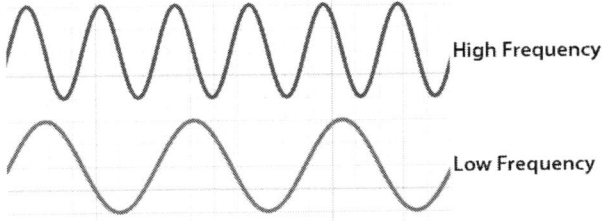

Now if you replace sound waves with light waves you will know why Hubble deduced that the universe was expanding. We already know that light is actually made up of different colors – red is lowest frequency and violet/blue is highest frequency.

The galaxies that Hubble observed were all reddish. This meant only one thing that the galaxies were going away from us. If galaxies could make sounds (just like cars) we could easily tell that they were going away! But astronomers study light, they have to rely on their eyes! And Edwin Hubble had discovered the most important fact of the 20th century – it was 1930.

Summary

By observing starlight, we can tell whether the

universe is expanding or contracting. Our universe is expanding and this was first realized by Edwin Hubble.

Applications

Fire Alarm uses photoelectric effect (particle nature of light)

Automatic doors – A person entering the room blocks the path of light and again, this is based on photoelectric effect (particle nature of light)

Solar Panel uses the principle of conservation of energy and an appropriate metal will allow maximum conversion. The metal for solar panels is chosen by taking photoelectric effect into account. Certain metals like Sodium are more effective than others like Silver

Night Vision Camera also uses photoelectric effect to view objects in the dark. This is shown in the figure below:

X-rays, microwave ovens, radio and radar navigation make use of electromagnetic radiation

Stars produce light because of conversion of mass into radiant energy. All stars are made up of hydrogen which is used up as fuel to ignite them

CRT TV uses an electron gun – a stream of electrons – to produce pictures on a fluorescent screen. When there is a magnet around a CRT TV, electron beam gets affected and thus the picture also gets fuzzy

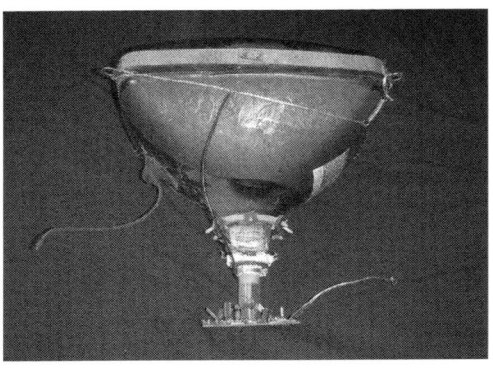

Look how the CRT picture gets affected by a magnet – below:

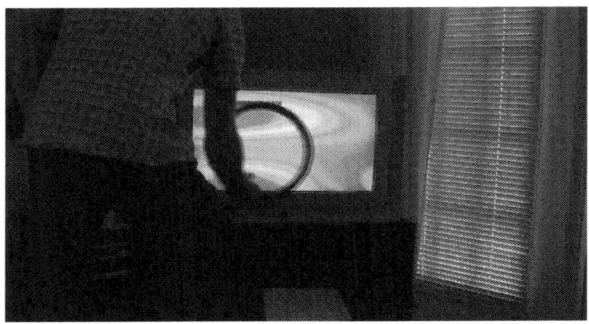

Atomic weapons make use of energy-mass equivalence. This is the famous equation by Albert Einstein. Atomic weapons can be very dangerous for humanity as they are weapons of mass destruction

Electron microscope uses wave-particle duality of electron. These microscopes have excellent magnification power!

Nuclear reactors are places where electricity is made from nuclear energy. Nuclear energy is made from mass of a heavy nucleus such as Uranium.

Einstein's general relativity theory says that gravity curves space and time, resulting in a tendency for the orbiting clocks to run slightly faster, by about 45 microseconds per day. The net result is that time on a **GPS** clock runs faster than a clock on the ground! This has to be corrected by taking into account the effects of Einstein's **relativity!** GPS has to be very accurate for proper functioning...

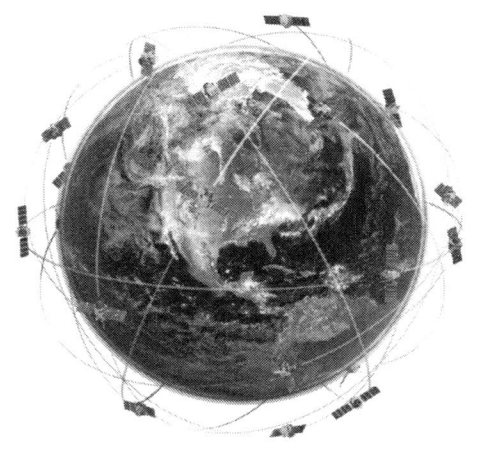

GPS helps in navigation – say thank you to Einstein!

Questions

1. Why was Newton particularly interested in the particle theory of light?
2. If an object has mass 10 kg and speed 2 m/s what will be its momentum? If "matter wavelength" is given as **h/momentum** then what will be its wavelength? Here **h** is Planck's constant.
3. Why is wave nature of matter insignificant for massive objects?
4. What happens in a particle-particle collision?
5. Which color among the 7 colors of rainbow has the highest frequency? Which has the highest wavelength?

6. Arrange the following waves in increasing order of frequencies – visible light, radio waves, x-rays, micro waves
7. What is the electric charge of 500 electrons? Look up the electric charge of an electron on the web
8. If a photon of twice the energy hit an electron, with what energy will the electron come out of the metal?
9. What is the difference between constructive and destructive interference?
10. What is energy-mass equivalence? What will be the energy locked up in an electron? Look up the mass of electron on the web
11. Why did Louis Broglie propose the duality of matter?
12. How can light plus light create darkness? What is this

called? And which theory can explain this phenomenon?
13. Why was wave-particle duality accepted in the case of light?
14. Which is the heaviest sub-atomic particle among electron, proton and neutron?
15. What is plum-pudding model? Why was it inappropriate?
16. What is an alpha particle?
17. How was the alpha particle scattering experiment performed? What were the outcomes of that experiment?
18. If an electron and proton are separated by distance of 1 meter and a neutron and proton are separated by the same distance in another universe. Then in which universe gravitational force is stronger? And in which

universe electric force is stronger?

19. How did Oersted discover the relation between electricity and magnetism?

20. What is an electromagnetic wave? Write the uses of the following – x-rays, radio waves.

21. If a charged particle is in acceleration mode what will it emit?

22. A static charged particle has its own electric field. What does a moving charged particle have?

23. In a given atom there are 4 protons and 4 electrons. This atom is stable electrically. What can you say about the charge of electron with respect to the charge on proton?

24. What was the problem in Rutherford's model of atom?

25. How did Bohr correct the problem in Rutherford's model of atom?
26. If a nucleus is the tip of a pin then the outermost region of the atom would be a meter away. Use scaling method for the following – if the nucleus is the size of an orange, where will the outermost part of the atom be? Use your own dimensions on assumption.
27. Despite having positive charged particles in a very small space, a nucleus is still stabilized by what kind of energy?
28. By observing the binding energy curve, which of the following has higher binding energy – Copper or Iron?
29. What will happen if a uranium atom is hit by speeding particles?

30. Differentiate between fusion and fission. Where are these used?
31. If a car is moving at the speed of 40 m/s and a bike opposite it moves at 5 m/s. what will be the speed of the car as observed by the bike rider?
32. What are the two postulates of special relativity?
33. What is a frame of reference? What is absolute rest?
34. If different observers in space measure the speed of light, what will they get?
35. What was Einstein's idea about gravity? How was it different from Newton's idea?
36. What are gravitational waves? How were they discovered recently?
37. Electric force is attractive as well as repulsive. On the

other hand gravitational force is only attractive. What other differences can you think of between these two forces?

38. What is general relativity? What does it unify?

39. What is a time machine? Is it possible to build a time machine?

40. What is a black hole?

41. What information can be derived from observing stars? How is that useful?

42. What is redshift?

43. What is Doppler Effect in the case of sound waves? How is it different from that in the case of light waves?

44. What did Edwin Hubble observe when he studied the lights of different galaxies?

45. What did Hubble conclude after making his observation? How did Hubble's

observation affect Einstein's dream?
46. If the universe is expanding what will we get if we reverse the *universe movie?* Will it start contracting? And to what end?

Glossary

Wave-particle duality- The concept that wave and particle nature are simultaneous
Photoelectric Effect- Emission of electron when light is incident on a metal
Momentum- The product of mass and velocity of an object
Quantum- A packet of discrete energy
Sub-atomic- Smaller than an atom
Alpha particle- A positive particle
Energy state- Fixed energy level in which an electron can be found
Uncertainty principle- Finding position and velocity of a particle is not possible simultaneously
Nucleus- Dense part of an atom
Nuclear force- A force stronger than electromagnetism
Electric charge- A property of a particle
Binding energy- Responsible for holding the nucleus together
Radioactive decay- Emission of particles out of the nucleus

Energy-mass equivalence- Mass is compact form o energy
Frame of reference- A matter of perspective
Relativity- Relative measurement
Cosmic speed limit- Speed of light
Gravitational waves- Ripples of gravitational force
Time machine- A machine capable of manipulating time
Black hole- A dead star with very strong gravitational field
Doppler Effect- Relative change of frequency of source of waves
Static Universe- Unchanging universe
Expanding universe- Changing universe
Redshift- Starlight appearing reddish

About the Author

Vedang Sati *(Illustrator, Physics Enthusiast)* is the founder of WonderPHY6, an online science community primarily interested in the promotion of physics to younger audiences. He was awarded Student of the Year award (2013) by The Times of India Newspaper in Education (NiE) for excellent academic and non-academic performance.

Vedang started working as a science promoter when he was only 16. And to interact with more number of people, he set up an online medium called WonderPHY6 in 2014. As of today, WonderPHY6 has a weekly reach of over 70,000 unique people (source - Facebook Analytics). Currently, he is an Engineering undergraduate student studying the rich fundamentals of electronics.

Curious?
Visit **www.facebook.com/wonderphy6**
If you liked this book, please do write a review on
www.amazon.com/Modern-Physics-Beginners-Relativity-Universe-ebook/dp/B00LH0GREK

Thank You

WonderPHY6

WonderPHY6 is an online community primarily interested in the promotion of physics to younger audiences. A group of enthusiastic engineering and art students dedicated to make physics entertaining. Wonderphy6 has a weekly reach of over 60,000 unique people and the number is just growing every week.

Printed in Great Britain
by Amazon